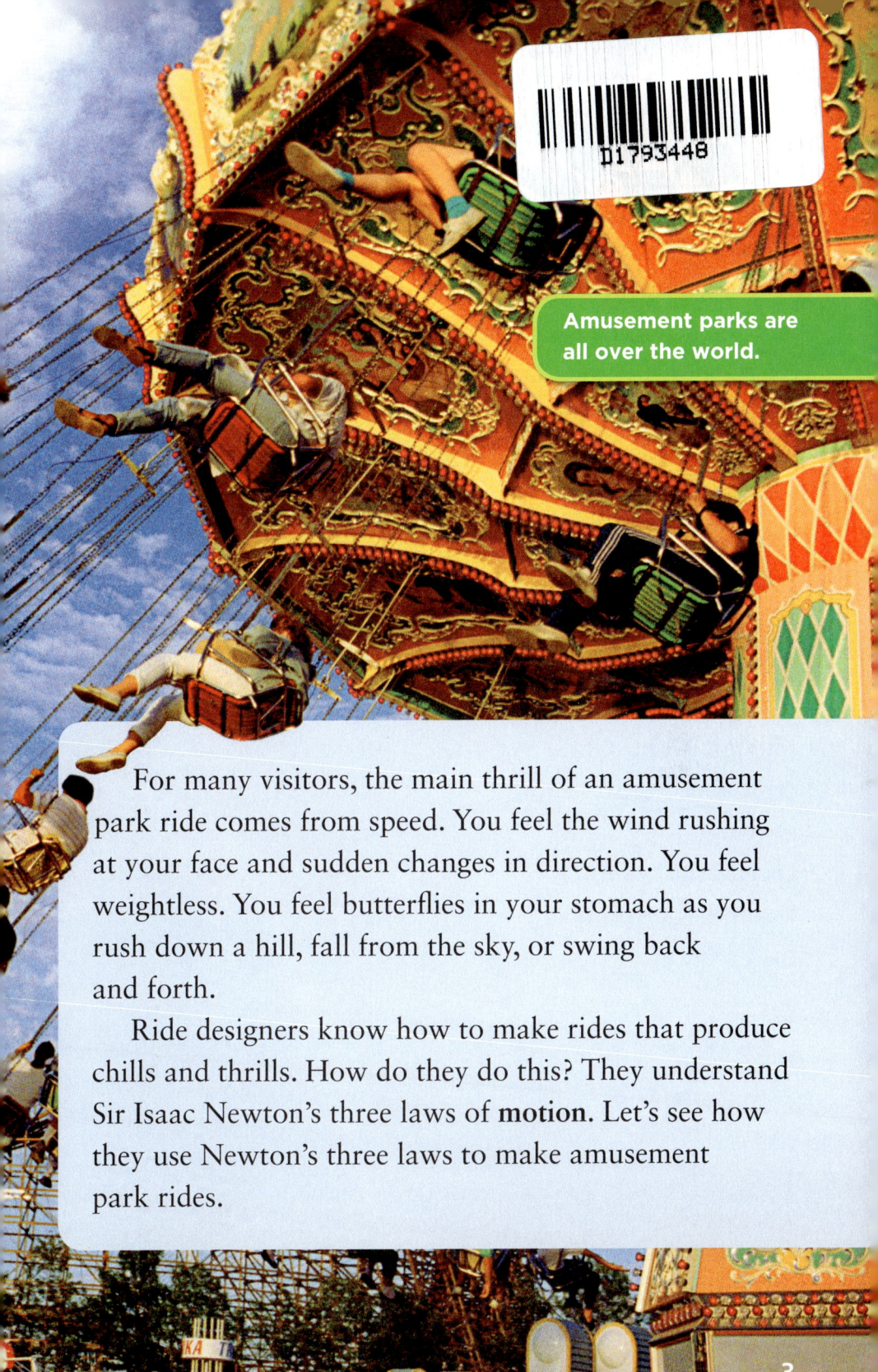

Amusement parks are all over the world.

For many visitors, the main thrill of an amusement park ride comes from speed. You feel the wind rushing at your face and sudden changes in direction. You feel weightless. You feel butterflies in your stomach as you rush down a hill, fall from the sky, or swing back and forth.

Ride designers know how to make rides that produce chills and thrills. How do they do this? They understand Sir Isaac Newton's three laws of **motion**. Let's see how they use Newton's three laws to make amusement park rides.

Chapter 2

Newton's Three Laws of Motion

Sir Isaac Newton was a genius. He understood math and science like no one before him. He developed the idea of the three laws of motion. Let's see how Newton's laws (or rules) are used to design amusement park rides.

Newton's first law of motion says that an object at rest will stay at rest unless a **force** acts upon it and makes it move. A moving object keeps moving in the same direction. It also keeps moving at the same speed unless a force changes the speed or direction.

Sir Isaac Newton
(1643-1727)

Amusement Park Rides

by BARBARA M. LINDE

Table of Contents

Chapter 1
Welcome to the Amusement Park . . 2

Chapter 2
Newton's Three Laws of Motion . . . 4

Chapter 3
Forces 6

Chapter 4
Roller Coasters 8

Chapter 5
Bumper Cars 14

Chapter 6
A Swinging Pendulum Ride 20

Chapter 7
The Aerial Tramway 24

Chapter 8
Meet a Roller Coaster Designer . . 28

Glossary 31

Index 32

Chapter 1

Welcome to the Amusement Park

You hear a roaring sound and then ear-splitting screams. You see speeding cars climbing long ramps, zipping through loops, and gliding smoothly to a stop. You feel the splash of hundreds of liters of water as you go down a watery ramp. Where are you? What is happening? You are enjoying a day at an amusement park!

Many people enjoy the thrill of amusement park rides.

This ride is at rest and will stay at rest unless a force makes it move.

Newton's second law of motion states that the change in motion of an object depends on the amount of force used on the object. The change in motion also depends on the mass of the object. A small force can change the motion of an object with little mass. A larger force is needed to quickly change the motion of an object with more mass.

Newton's third law says that for every action, there is an equal and opposite reaction. One object applies a force to a second object. The first object also feels a force. Yet it is in the opposite direction of the applied force.

Chapter 3
Forces

There is friction between your foot and the basketball court.

Designers of amusement park rides use Newton's laws of force and motion. Force and motion provide the thrill of a theme park ride! Here's how.

You feel forces every day. When you lift a heavy object, you work against **gravity**. Gravity is the force that pulls objects toward Earth.

When two objects rub together, they produce a force called **friction**. When you change directions on a basketball court, you rely on the friction between your sneaker and the floor.

Moving through air or water produces a force like friction. This force is called **drag**. When you run, you can feel the drag from the air on your body.

Scientists measure some forces relative to the force of gravity. These are called **g-forces**. The normal force of gravity on Earth is 1g. Astronauts experience weightlessness, or 0 g's. A force of 10 g's is ten times the force of gravity. You experience changes in g-forces on a roller coaster.

When you are on a roller coaster, you can feel the changes in the g-forces.

Chapter 4
Roller Coasters

This roller coaster train is at rest in the station.

Now you know about forces and motion. Let's look at some amusement park rides that use Newton's laws.

Suppose you have just climbed into a car of a roller coaster train. The train is at rest on the track. According to Newton's laws, a force is needed to start the train in motion. Otherwise, it will stay at rest. After you are safely strapped in, the ride operator starts the ride.

An electric motor turns a wheel. Using friction, the rubber tire on the wheel pushes the train out of the station.

After the cars leave the station, they continue to roll smoothly on the track. The track is usually flat for this part of the ride. There are no new forces on the cars. This is the second part of Newton's first law.

Now you hear the clack of the chain under the cars. You wonder, "What is going to happen next?"

This roller coaster keeps going once it is in motion!

The chain pulls the roller coaster up the first hill.

A powerful electric motor turns a **pulley**. The pulley applies force to a chain. A latch at the bottom of the train connects the chain. It then pulls the roller coaster train up the first hill. The force of gravity is trying to make the train roll back down the hill. But the chain has tension. This tension is the force that pulls the weight of the train and the passengers inside it.

Now the train is at the top of the first hill. You take a deep breath. You know what is going to happen. As the cars pass over the hill, gravity takes over. The plunge begins. Down you go! During the downhill ride, the force of gravity continues to act. The longer the force acts, the faster you go. This is Newton's second law of motion. The longer the first hill, the more speed the train gains.

Hold on tight!

The first hill is a big thrill!

At the bottom of the hill, the track changes direction. It goes up again. Newton's second law says that a force is needed to produce this change in direction. You experience the g-force as the track pushes the car upward.

Suddenly, the track curves to the right. You are thrown against the left side of the car. But your body continues to move in a straight line. It will continue until the forces on your left make you move in the same direction as the car. This is Newton's first law in action.

There are many curves on a roller coaster ride.

Finally, you come to the end of the ride. Powerful brakes clamp onto the bottom of the car. The force of the friction between the clamp and the car brings the car to a stop. Your safety harness releases you. You walk from the car on shaky legs.

What a ride! Now where are the bumper cars?

Whew! What a ride!

Chapter 5
Bumper Cars

Like a regular car, a bumper car has a steering wheel, a floor pedal, and wheels. The wheels are hidden under the big bumper. There is a pole at the back of the car. It has electric wires in it.

The electrical grid stretches across the ceiling of the arena.

Bumper cars use electricity to move.

The top of the pole makes contact with an electrical grid. The electricity travels down the pole to an electric motor in the car. The motor moves the car across the floor. You press on a floor pedal to start the motor.

The electric motor turns a wheel. The force of friction between that wheel and the floor starts the car in motion. This is Newton's first law in action. It says that an object at rest stays at rest unless a force acts upon it.

As the driver, you control the car's direction with the steering wheel. The steering wheel is attached to a column. The column is attached to a wheel under the car that touches the floor. Friction controls which way the car moves. If there weren't any friction between the wheel and the floor, you could turn the steering wheel over and over again, but the direction would not change.

When you were bumped from the left, your car went to the right. The car that bumped you bounced back to the left.

Wham! A car bumps you from the left. The force of the bump moves your car to the right. The other driver has also felt the bump. That car bounced back toward the left. This is Newton's third law of motion. It says that if one object exerts a force on another object, that object exerts the same amount of force right back.

Get ready to bump and be bumped!

Did you notice that the floor was slippery? The slippery floor produces only a small amount of friction with the wheels. The amount of force from the friction is smaller than the force from a collision with another car. This makes the car move sideways when it is bumped.

> The floor of the arena is flat and slippery. Be careful!

You must wear a seat belt in a bumper car.

Bam! You just hit the back of a car. The force on the bumper of your car slows the car down. Yet your body continues moving forward until the seat belt stops you. Newton's first law says that an object in motion (you) stays in motion until it is acted on by an outside force (the seat belt).

When driving bumper cars, you experience Newton's laws over and over again. Your ride is over when the operator turns off the electricity.

Don't worry, the fun is not over. It is time for the **pendulum** ride.

Chapter 6
A Swinging Pendulum Ride

A pendulum ride swings back and forth.

The car on the pendulum ride looks like the curved part of a boat. About 40 people can sit in it. Usually, there are rows of benches, with four people on a bench. The car hangs from a support structure. It looks like a large playground swing. This giant structure can be about 16 meters (52 feet) tall!

You climb into the car, sit down, and pull the safety bar down into your lap. Then the operator starts the ride.

A motor pushes the car forward and upward. As you go up, you feel the back of the bench supporting your weight. The bench experiences a force because of your weight. This is Newton's third law of motion. It says that for every action, there is an equal and opposite reaction.

The pendulum ride has a strong support structure.

The car reaches the top of its swing. Then it goes backward. Both you and the car are falling back toward Earth. There is a sinking feeling in your stomach. This is because there are fewer g-forces now.

Newton's first law tells you that when the car reaches the bottom of its swing, it will keep moving in a straight line. But this doesn't happen. Why? The support structure applies an upward force to the car. Newton's second law says that a force produces a change in speed or direction. You feel that force in addition to the usual force of gravity. This makes you feel heavier.

You'll get a sinking feeling in your stomach on this ride!

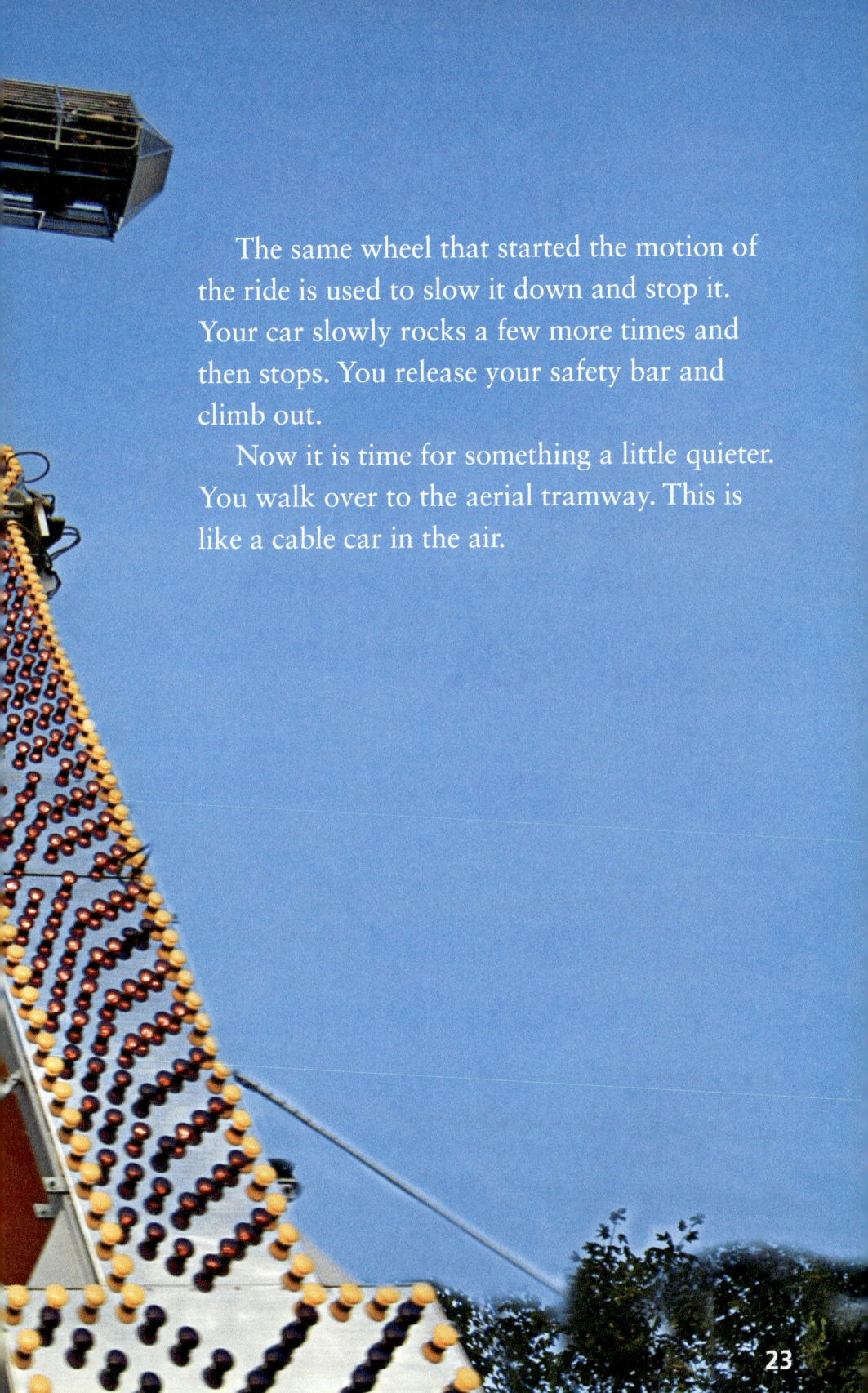

The same wheel that started the motion of the ride is used to slow it down and stop it. Your car slowly rocks a few more times and then stops. You release your safety bar and climb out.

Now it is time for something a little quieter. You walk over to the aerial tramway. This is like a cable car in the air.

Chapter 7

The Aerial Tramway

You are standing in line at the aerial tramway station. This ride will take you high above the treetops.

There are two stations, one at each end of the ride. There is a motor with a pulley at one station. There is a pulley at the other station. A heavy steel cable travels in a loop around the pulleys and over rollers on tall poles. Cars hang from arms. The arms are attached to the cable.

Cars on the aerial tramway travel on cables.

As the car comes to the station, it goes from the cable onto a side track. Now there is no outside force on the car. Yet it still keeps moving, following the side track. The car stays in motion until the ride attendant grabs it to stop it. Newton's first law happens again!

You are about 12 meters (40 feet) above the park!

The ride attendant lets you into the car. The car travels down the sloping track. The track meets with the cable. The car attaches to the cable. This happens smoothly because there is no change in speed or direction.

The aerial tramway is not like the other rides. This ride is smooth and steady. Your body will not feel extra g-forces or bumps. Since you are up so high, you can look down at the other rides. Think about all the forces that those rides produce.

Enjoy the smooth, gentle ride.

After riding the aerial tramway, it is time to leave. Think about your fun-filled day. You now know more about how amusement park rides work. You have experienced Newton's three laws of motion. You cannot wait to tell everyone all about the rides!

The aerial tramway gives you a view of the entire park.

Chapter 8

Meet a Roller Coaster Designer

Jim Seay designs roller coasters.

Jim Seay is the president of a company that designs and builds rides for amusement parks. Jim answered some questions about his job as a roller coaster designer.

▶ **How did you get interested in designing roller coasters?**
I used to work for an aircraft company. I also raced sailboats. While racing, I met people in the theme park industry. I learned that people use the same concepts of engineering and design for amusement park rides that they do for aircrafts.

▶ **What should someone who wants to design roller coasters study in school?**

He or she should study mathematics and science.

▶ **Which drop is the highest drop and why?**

Each year, the highest drop gets higher and higher. Now there is equipment that goes over [122 meters] 400 feet. My company is planning drops above [152 meters] 500 feet. The heights keep getting higher because every [amusement park] wants to say it has the tallest roller coaster.

▶ **What is the fastest speed you have ever designed for a coaster?**

We have designed a ride that goes faster than [320 kilometers] 200 miles per hour. But we haven't built it yet.

▶ **About how long does it take to build a roller coaster?**

It takes about two years to design and build a roller coaster.

▶ **What are the most rewarding parts of your job?**

I enjoy seeing the excitement and satisfaction on the riders' faces.

Roller coasters are exciting rides!

Glossary

drag (DRAG) friction between an object and air or water *(page 7)*

force (FORS) a push or a pull that makes something move or change its direction *(page 4)*

friction (FRIK-shuhn) rubbing of one object against another; a force that slows down moving things *(page 6)*

g-force (GEE-fors) a force measured relative to the force of gravity *(page 7)*

gravity (GRAV-i-tee) a force that pulls objects toward the center of Earth *(page 6)*

motion (MOH-shuhn) movement; a change in place or position *(page 3)*

pendulum (PEN-juh-luhm) a weight that hangs from a fixed point and swings back and forth because of gravity *(page 19)*

pulley (PUL-ee) a wheel with a groove around it that a rope or chain can be pulled over to lift things; a wheel pulled by or driving a belt *(page 10)*

Index

aerial tramway, 23-27

amusement park, 2-3, 6, 8

bumper car, 13-19

direction, 3-4, 6, 12, 16, 22, 26

drag, 7

force, 4-13, 15, 17-19, 21-22, 25-26

friction, 6-7, 13, 15-16, 18

g-force, 7, 26

gravity, 6-7, 10-11, 22

laws of motion, 3-6, 8-9, 11-12, 15, 17, 19, 21-22, 25, 27

mass, 5

motion, 3-6, 8, 15, 19, 23, 25, 27

motor, 9-10, 15, 21, 24

Newton, Sir Isaac, 3-6, 8-9, 11-12, 15, 17, 19, 21-22, 25, 27

pendulum ride, 20-21

pulley, 10, 24

roller coaster, 7-13, 28-30

Seay, Jim, 28

speed, 3, 11, 22

weightlessness, 3, 7